WILDLIFE IN BLOOM SERIES

Little Bear

BY AUTHOR & CONSERVATIONIST

LINDA BLACKMOOR

ISBN: 979-8-9904465-8-8 (PRINT)

PUBLISHED BY QUILL PRESS. LINDA BLACKMOOR'S TITLES MAY BE
PURCHASED IN BULK FOR EDUCATIONAL, BUSINESS, FUNDRAISING, OR
SALES PROMOTIONAL USE. FOR INFORMATION, PLEASE EMAIL
HELLO@LINDABLACKMOOR.COM

FIRST PRINT EDITION: 2024

LINDA BLACKMOOR
WWW.LINDABLACKMOOR.COM

SPECIES

Bears include eight remarkable species across the globe: the brown bear, American black bear, polar bear, Asiatic black bear, sloth bear, sun bear, spectacled bear, and the giant panda. The brown bear, which includes the grizzly bear, roams forests and mountains in North America, Europe, and Asia. Polar bears tread the icy Arctic, while giant pandas dwell in China's bamboo forests.

HIBERNATE

Some bears enter a deep sleep called hibernation during winter months to survive when food is scarce. During hibernation, a bear's heart rate drops from 40 beats per minute to as few as 8, and they can go months without eating, drinking, or eliminating waste. Pregnant females give birth during this time. This adaptation allows bears to conserve energy and endure harsh climates.

OMNIVORE

Bears are omnivores with diets ranging from berries and nuts to fish and small mammals, depending on species and season. The giant panda's diet is unique, consisting almost entirely of bamboo—eating up to 40 pounds daily. In contrast, polar bears are mostly carnivorous, relying on seals as their primary food source in the Arctic. This diet flexibility demonstrates bears' adaptability to different habitats.

SCENT

Bears possess an extraordinary sense of smell, believed to be the strongest of any land mammal. A polar bear can detect a seal's breathing hole under three feet of compacted snow from over a mile away. Their olfactory abilities help them locate food, recognize other bears, and navigate vast territories. This heightened sense is vital for their survival in diverse environments.

SMART

Bears exhibit remarkable intelligence and problem-solving skills, capable of using tools and remembering locations. They've been observed rolling rocks to access food or using sticks to scratch themselves. Bears have excellent memories, recalling food sources and navigation routes over large areas. Their cognitive abilities reflect a high level of adaptability and learning.

SIGNALS

Bears communicate through vocalizations, body language, and scent markings. They make sounds like grunts, roars, and huffs to express emotions or signal other bears. Claw marks on trees and rubbing against surfaces leave scent cues for mating or marking territory. This complex communication system helps them interact and coexist within their habitats.

SIZE

Bears are among the largest land mammals, with polar bears and Kodiak bears reaching up to 10 feet tall when standing and weighing over 1,500 pounds. The smallest species, the sun bear, stands about 4 feet tall and weighs up to 150 pounds. This size variation reflects adaptation to different environments and lifestyles. Their powerful physiques enable them to be dominant predators and foragers.

HABITAT

Bears inhabit a wide range of environments—from Arctic ice fields to tropical rainforests. Polar bears rely on sea ice for hunting seals, while spectacled bears dwell in the cloud forests of the Andes Mountains. Their adaptability to diverse habitats showcases resilience and ecological importance. Bears play crucial roles in ecosystems, such as seed dispersal and regulating prey populations.

BEAR FACT #9

CUBS

Bear cubs are typically born during hibernation, blind and hairless, weighing less than a pound. A mother bear is fiercely protective, nurturing her cubs for up to two years and teaching them survival skills. Cubs learn to climb trees, forage for food, and navigate their surroundings under her guidance. This maternal care is vital for their development and future independence.

SOLITUDE

Most bear species are solitary creatures, coming together only during mating season or when food is abundant. Their solitary nature reduces competition for resources within their territories. However, they remain aware of other bears through scent markings and vocalizations. This behavior reflects a balance between independence and necessary communication.

THREATS

Many bear species face threats from habitat loss, climate change, and poaching, leading to declining populations. The polar bear is classified as vulnerable due to melting Arctic ice, while the giant panda was recently upgraded from endangered to vulnerable thanks to conservation efforts. Protecting bears involves preserving habitats and reducing human-wildlife conflicts.

SWIMMING

Bears are excellent swimmers, especially polar bears, who can swim for days and cover hundreds of miles across Arctic waters. Their large paws act like paddles, and layers of fat provide buoyancy and insulation. American black bears and brown bears also swim well, crossing rivers and lakes in search of food. Swimming skills are essential for accessing resources and migrating between habitats.

BEAR FACT #13

CLIMBERS

Many bear species are adept climbers, using strong claws to ascend trees. American black bears often climb to escape danger or find food like nuts and fruits. Sun bears have especially long claws for climbing in tropical forests. Climbing allows bears to exploit various food sources and avoid ground–level threats.